Dedicated to Water

THE ART OF JOY
PUBLICATIONS

Published by The Art of Joy Publications

Email: theartofjoypublications@gmail.com

First Published 2017

ISBN 978-0-9956110-5-4

Where does our drinking water come from?
How does it reach us and how can we safeguard it?

These are very important questions that we should all be asking because fresh water is essential for every aspect of our lives and yet there is a surprising lack of awareness about how the global water cycle works and how we can safeguard it. For our long-term survival we all need to become much better informed about fresh water so that we can do what is needed to make sure that it continues to circulate around the world and get to the places where it is needed.

Freshwater is circulated around the Earth via the water cycle, also known as the hydrological cycle. This cycle is dependent upon healthily functioning ecosystems such as mountains, glaciers, mountain forests, rain forests, wetlands, rivers, aquifers, lakes and oceans. Many of these ecosystems have been dramatically degraded and this is destabilising the global water cycle. Greater general awareness about this situation is needed so that appropriate action can be taken to rectify it.

This book is intended to provide food for thought highlighting key points that need considering, pertinent quotes from experts, problems that need resolving and innovative solutions that can be applied.

Water is the most abundant and precious substance on the surface of planet Earth. Yet how much do we really understand its Nature and qualities?

We live on a planet upon which the surface is largely composed of water. We humans and all other life forms are also largely composed of water. This is the nature of life on planet Earth.

Water is a mystery and a beauty and we all know we need it to live. Yet how much do we know about it reaching Earth's rivers and lakes or even our taps?

Water is unusual, as it exists in three distinct states and is constantly fluctuating between these states. The first of these is an invisible gas known as water vapour. We only see this when it condenses and forms clouds.

In its second state water is a solid and is called ice and snow. In these forms it can be stored for long periods. This is why the mountains of the world are also known of as Earth's fresh water storage tanks.

The third state of water is liquid. Fresh water runs in rivers, streams and springs, above and below the Earth from the mountains down to the seas.

"The cycling of water among the three phases is overwhelmingly important for Earth, driving not just the atmospheric general circulation, but also the very existence of life as we know it."

(USGCRP, Part 1 The Water Cycle Science Plan 2001)

Although only about 3% of the Earth's water is freshwater, due to the recycling function of the water cycle, it has been able to sustain life on Earth for millions of years.

The hydrological cycle is also the mechanism, which is at the heart of both the causes and the effects of climate change (USGCRP, Part 1 The Water Cycle Science Plan 2001). Climatic stability is dependent upon its balance.

Fresh water is considered to be a limited resource. However due to it's recycling and regenerating mechanism, it can provide enough for billions of beings over vast time scales. Hence it is does not decrease in the same manner as limited resources.

For the global water cycle to function in a balanced and effective manner, it is utterly dependent upon mixed forests with their huge variety of plant species and biodiversity.

"Forests play a pivotal role
in the hydrological cycle"

(UNEP, CBD, 'Water, Wetlands and Forests', 2010)

"the wide combinations of causes of water scarcity are all considered to be related to human interference with the water cycle." (FAO, Coping with Water Scarcity, 2008)

Monoculture forests do not perform all the life sustaining functions that biodiversity does. They merely exacerbate water and soil erosion related problems and thus desertification.

"Large-scale monoculture tree plantations have proven to cause a broad range of negative environmental and social impacts. They have been and continue to be a major direct cause of the destruction of native forests in countries as varied as Australia, Brazil, Indonesia, and Chile." (Global Forest Coalition, 2011.)

"Forests perform vital ecosystem services, including the regulation of the water and carbon cycles and protection of biodiversity, that are essential to food production and food security and nutrition in the long term."

(FAO/HLPE, Sustainable Forestry for Food Security and Nutrition, 2017)

"an ecosystem approach to water management
from local to continental levels is key to
ensuring quality and quantity of water for
food security and nutrition in the future."

(FAO/HLPE, Water for Food Security and Nutrition, 2015)

Recent studies into bio-precipitation are indicating that forests and biodiversity play a much larger role in precipitation and climate regulation, than previously imagined (Brent Christner, 2009).

Ice nucleating bacteria that live on the surface of plants, and biological particles in the atmosphere may play an essential role in seeding clouds (David Sands, 2008).

"Biological ice nucleators (IN)
are the most active IN in nature"

(Brent Christner et al., Science, 'Ubiquity of Biological
Ice Nucleators in Snowfall', 2008)

"We recognize the key role that ecosystems play in maintaining water quantity and quality"

(UN, The Future We Want, 2012)

Unfortunately many of the crucial ecosystems, which are essential for fresh water supply and regulation, have been massively depleted worldwide.

"Rivers that for centuries ran from source to sea now run dry in many years due to damming, diversion and depletion of glaciers and water resources."

(UN WATER, The Global Water Crisis, 2012)

"There is abundant evidence that changes in land cover and land use can have significant, even drastic, impacts on the water cycle at local and regional scales."

(USGCRP Report, A plan for a new science initiative on the global water cycle, 2001)

"Tropical deforestation, and the resultant effect on thunderstorm patterning, alters long-term weather patterns thousands of kilometres from the landscape disturbance."

(R.A. Pielke Sr, et al. Non Linearities in the Earth System, 2003)

"Each year an average of 13 million hectares of forest are removed."

(UN, New York Declaration on Forests, Climate Summit, 2014)

Earth's water cycle and climate systems are fundamentally interlinked and they depend upon ecosystem connectivity. This connectivity is rapidly being broken.

The widespread depletion of essential water related ecosystems is having catastrophic impacts upon climate and the global hydological cycle.

"The magnitude of the global freshwater crisis and the risks associated with it, have been greatly underestimated."

(UN WATER, The Global Water Crisis, 2012)

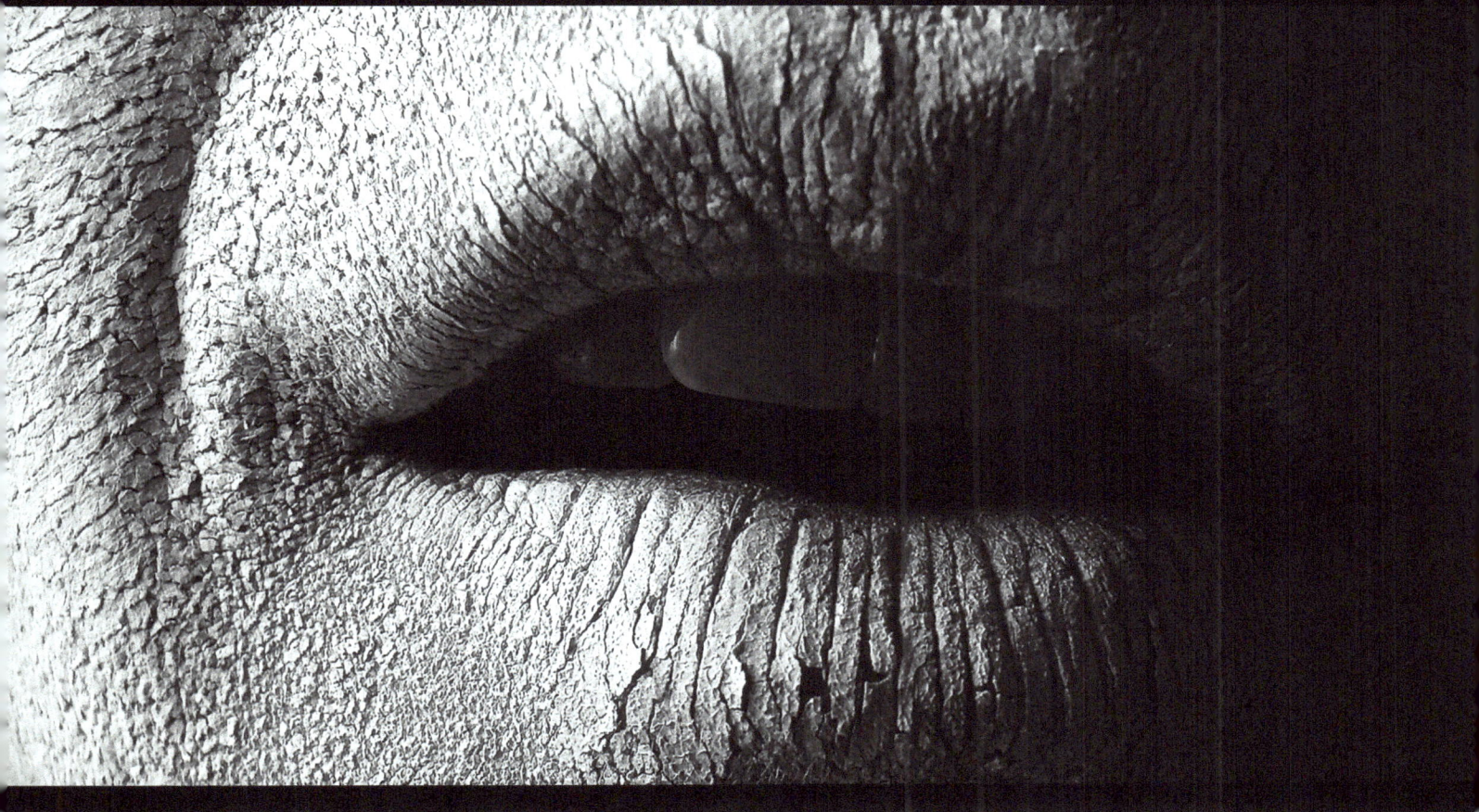

Our drinking water is threatened.

"By 2030 nearly half the global population could be facing water scarcity, with demand outstripping supply by 40 per cent."

(UN Secretary General, 22/3/13)

There is no alternative to water.
There is no escape from thirst.

Our lives and the lives of future generations depend upon whether we take the fresh water situation seriously or not.

To safeguard the global water cycle,
ecosystem rehabilitation needs to become
a top priority for the global community.

We need to know where our water comes from, how it reaches us and make sure that we protect the global water cycle for future generations.

It is essential that more attention be given to protecting the key natural components of the global water cycle.

Mountain ecosystems play a pivotal role in regulating global water and climate systems. Yet they are particularly fragile and been have been intensively depleted. They therefore need special consideration.

"Mountains and uplands cover about 24 per cent of the Earth's surface, and influence most of the planet. The most important influence is on the hydrological cycle."

(Hans J.A. van Ginkel. Rector, UN University, Under-Secretary General, 2004)

"Mountain forests also occupy a crucial position in terms of climate change, representing fundamental ecosystems for the health of the planet."

(FAO, Mountain Forests in a Changing World, 2011)

Seasonal snow cover and glaciers store large amounts of freshwater and are therefore critical componentsof the land surface hydrological cycle."

(USGCRP. A plan for a new science initiative on the global water cycle, chapter 2, 2001)

"Mountain ecosystems such as mountain forests, cloud forests, wetlands and grasslands play vital roles in water storage and supply"

(UNESCO, Climate Change Impacts on Mountain Regions of the World, 2013)

"All of the world's rivers originate in the mountains and flow to the oceans, sustaining the life of all beings, in all ways of life here on Earth."

(United Nations, Agenda 21, 1992)

"Glaciers are a critical component of the Earth system and the present accelerated melting and retreat of glaciers has severe impacts on the environment and human well-being." (UNEP, WGMS, Global Glacier Changes, 2007)

"Given their important role in water supply and regulation, the protection, sustainable management and restoration of mountain ecosystems will be essential."

(UNESCO, Climate Change Impacts on Mountain Regions of the World, 2013)

"Today we are faced with a challenge that calls for a shift in our thinking, so that humanity stops threatening its life-support system."

(Wangari Maathai's Nobel Prize acceptance speech, 2004)

Integrated methods for safeguarding the global water cycle are needed and must be swiftly applied.

The Sacred Groves, Green Corridors Method
Creating a Green Biodiversity Network
To Preserve Fresh Water

We at Active Remedy Ltd. have given these issues a great deal of thought. Therefore we have designed an innovative program, which can be applied and adapted to make ecological restoration and preservation efforts far more effective.

This program, which has been designed in relation with mountain people of northern India, is known of as 'The Sacred Groves and Green Corridors' (SGGC) Method.

The Sacred Groves, Green Corridors (SGGC) method was created as a solution that can be practically implemented in order to safeguard the global water cycle and thus all sources of fresh water.

Essentially the SGGC method is a way of creating a global network of interconnected, healthy ecosystems, through which biodiversity can flow freely.

The SGGC method is intended to be simple, yet respectful of the requirements of local people, as well as being highly adaptable for the varied conditions found in mountainous regions.

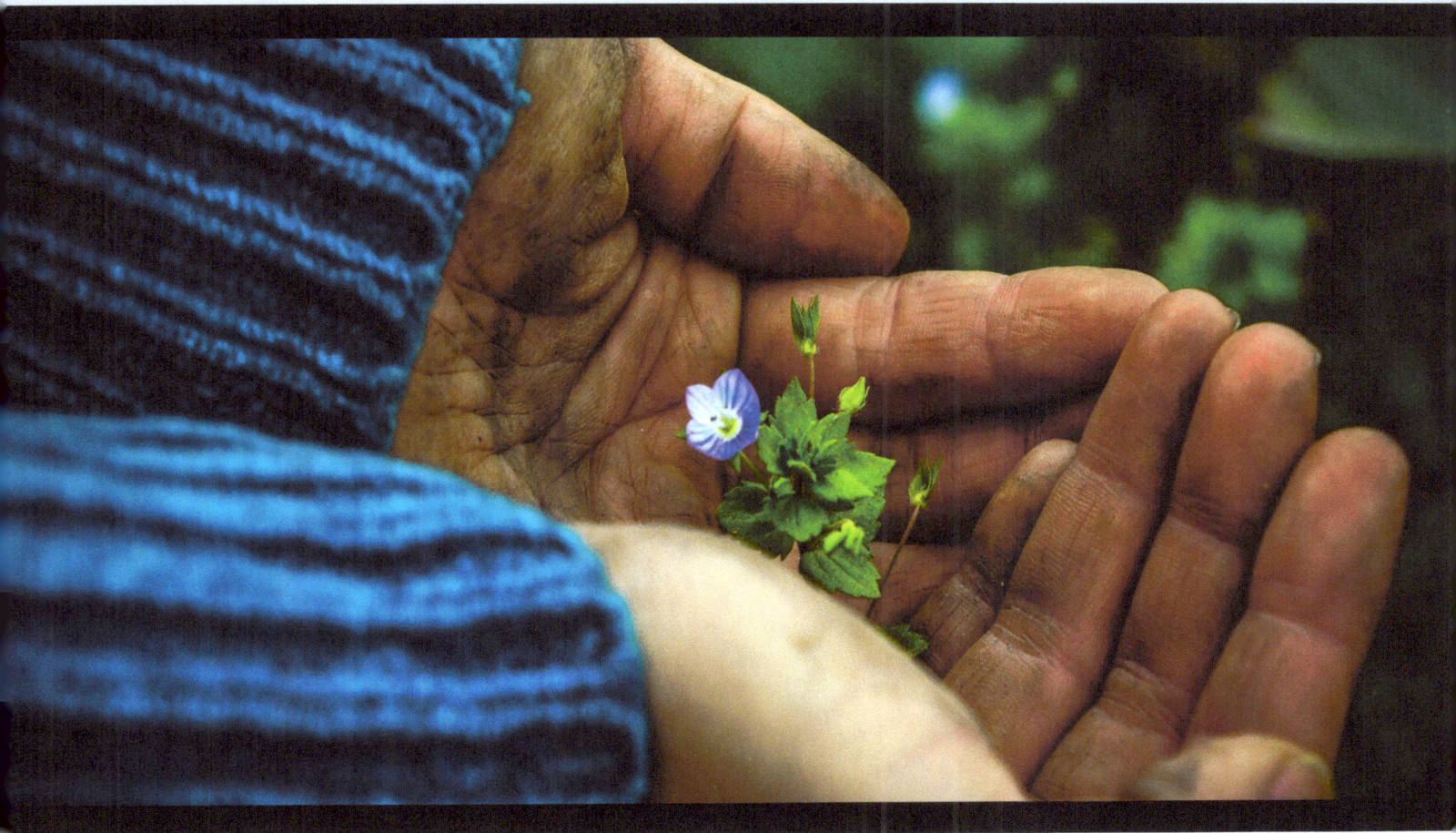

It is composed of a variety of approaches that are adaptable to the specific requirements of local ecosystems, values, spiritual customs and traditions.

Creating biodiversity corridors is an integral part of the SGGC method. These are an ideal and successful approach for efficiently spreading and linking biodiversity throughout mountain regions and many terrains.

These corridors would be environmentally restorative and support many of the needs of the local mountain communities by containing the various plants needed for fodder, fuel, medicines and also for local cottage industry.

These corridors can be grown and locally managed, using companion planting and Permaculture techniques, so that there would remain a constant presence of green plant cover, as well as a constant supply of natural resources that can be utilised by local communities.

Sacred groves are also a fundamental part of the SGGC method. Safeguarding and maintaining sacred groves is an ancient tradition that is recognised internationally and which has proven to be a highly effective in conserving biodiversity over thousands of years.

There are numerous sacred groves throughout the Himalayas but many of these are now isolated and vulnerable due to environmental and cultural degradation.

Linking these sacred groves with green biodiversity corridors would give greater shelter to the surrounding lands and help give protection against the intense effects of climatic fluctuations, droughts, landslides, floods and pests.

In order to protect our water, safeguarding the global water cycle needs to become a major global priority.

"Ensuring that ecosystems are protected and conserved is central to achieving water security – both for people and for nature. Ecosystems are vital to sustaining the quantity and quality of water available within a watershed, on which both nature and people rely."

(UN WATER, Water Security Analytical Brief, 2013)

Safeguarding freshwater for present and future generations is a matter of global Security!

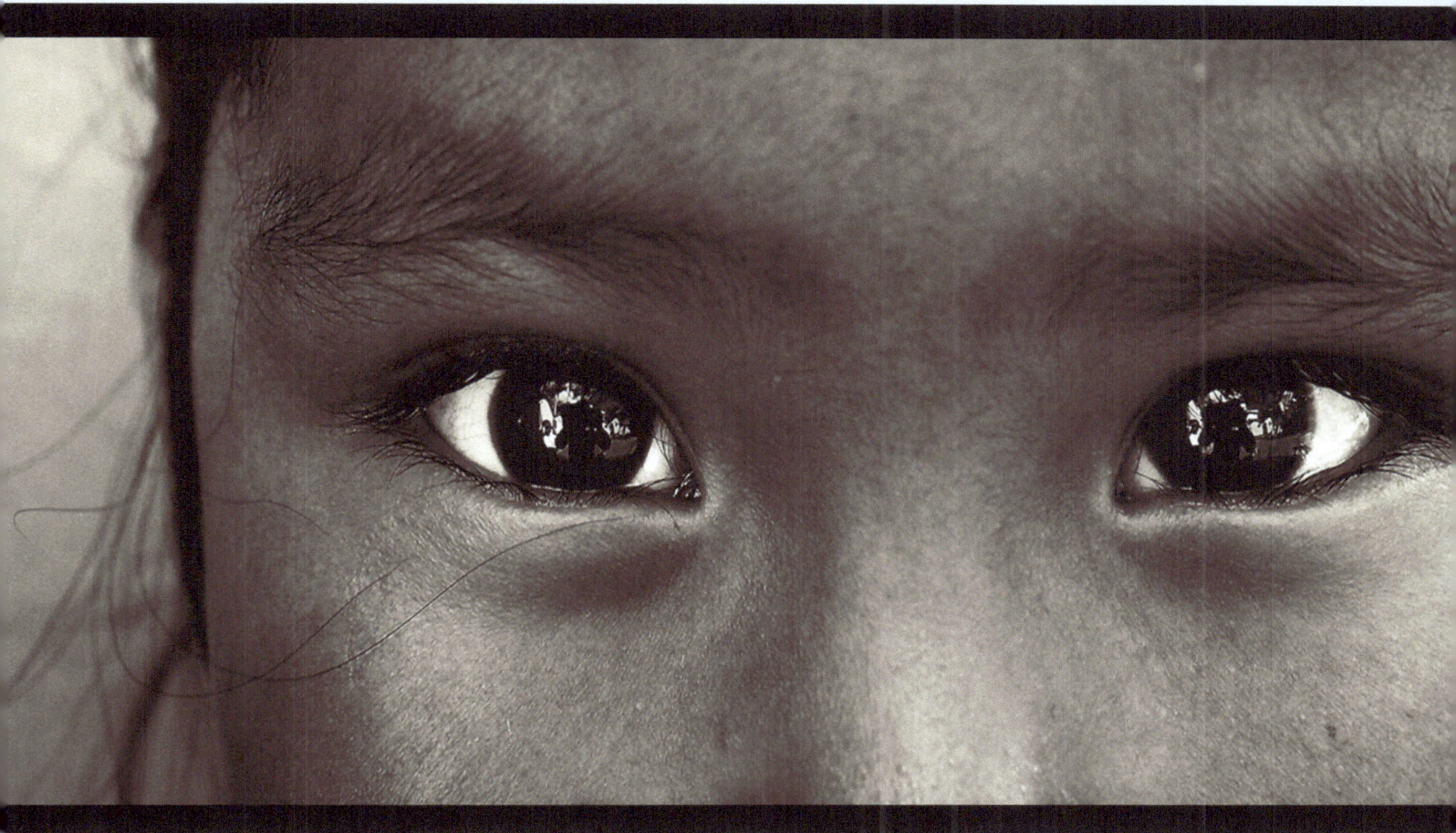

It is not a case of not being able to afford to deal with this, we cannot afford not to. Otherwise what will we say to the children?

It is recognised that fresh water and water security are central to global security and that these need to be included within all global defence strategies. Therefore at least 1% of the annual defence budget of all U.N countries should be allocated for maintaining the global water cycle.

If freshwater is treated simply as a commodity and safeguarding the global water cycle is ignored, we may soon be asking the question: "Who gets water and at what cost?"

In September 2015 in New York world governments signed the new Sustainable Development Goals. Target 6.6 States: "By 2020, protect and restore water-related ecosystems, including mountains, forests, wetlands, rivers, aquifers and lakes"

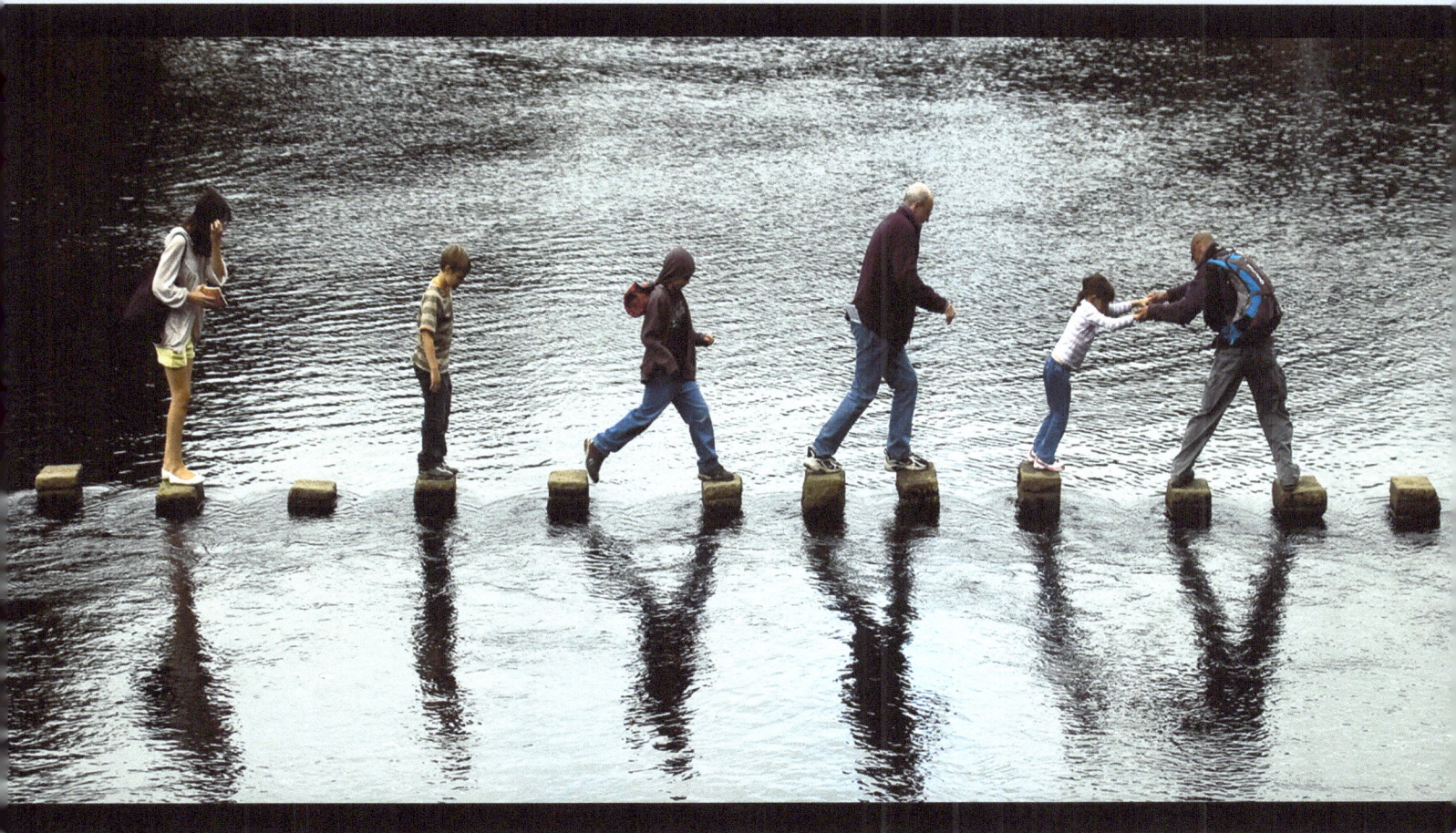

When we safeguard the water cycle we are safeguarding water, when we safeguard water we are safeguarding life and the well being of present and future generations and all life forms.

We decide our future

We, the global community, need to work together to safeguard our global water cycle.

Since 2005 we at Active Remedy Ltd. have been researching the relationship between the global water cycle, climate and the degradation of mixed mountain forests and essential ecosystems, which regulate these global systems.

Spurred by increasing instability in the climate systems and the global water cycle and the related increase of natural disasters, we at Active Remedy Ltd. have formulated an innovative method that can be applied and adapted to facilitate ecological restoration, preservation and adaptation efforts around world.

This method is called The Sacred Groves and Green Corridors (SGGC) method. The SGGC method has been formulated in conjunction with traditional indigenous mountain people over many years.

On the 30th of March 2016 the Secretariat of the United Nations Convention on Climate Change (UNFCCC) wrote to us inviting us to contribute towards a worldwide database on the use of Local, Indigenous and Traditional Knowledge and Practices for Climate Adaptation.
In response to this request, we submitted a report outlining the SGGC method. It can be found as a tool for implementation in their database.

To find out more about our work please visit our website:
www.activeremedy.org

ACTIVE R.E.M.E.D.Y LTD

We at Active Remedy feel deeply honoured that in May 2016 we were able to share the UNFCCC report with H.H the Dalai Lama in a private audience and that after taking the time to look through it and speak with us, he gave his blessings for the work of Active Remedy.

www.ingramcontent.com/pod-product-compliance
Lightning Source LLC
Chambersburg PA
CBHW061154030426
42336CB00002B/39